Alam Zeb, Reipar Nawaz

Effect of insect attack on the production of honey & honey bees in Nowshera District

M.Sc Forestry Research by Alamzeb Qazi

Der GRIN Verlag publiziert seit 1998 wissenschaftliche Arbeiten von Studenten, Hochschullehrern und anderen Akademikern als eBook und gedrucktes Buch. Die Verlagswebsite www.grin.com ist die ideale Plattform zur Veröffentlichung von Hausarbeiten, Abschlussarbeiten, wissenschaftlichen Aufsätzen, Dissertationen und Fachbüchern.

Document Nr. V201204

Alam Zeb, Reipar Nawaz

Effect of insect attack on the production of honey & honey bees in Nowshera District

M.Sc Forestry Research by Alamzeb Qazi

GRIN Verlag

Die Deutsche Bibliothek verzeichnet diese Publikation in der Deutschen Nationalbibliografie; detaillierte bibliografische Daten sind im Internet über http://dnb.d-nb.de/ abrufbar.

Dieses Werk sowie alle darin enthaltenen einzelnen Beiträge und Abbildungen sind urheberrechtlich geschützt. Jede Verwertung, die nicht ausdrücklich vom Urheberrechtsschutz zugelassen ist, bedarf der vorherigen Zustimmung des Verlages. Das gilt insbesondere für Vervielfältigungen, Bearbeitungen, Übersetzungen, Mikroverfilmungen, Auswertungen durch Datenbanken und für die Einspeicherung und Verarbeitung in elektronische Systeme. Alle Rechte, auch die des auszugsweisen Nachdrucks, der fotomechanischen Wiedergabe (einschließlich Mikrokopie) sowie der Auswertung durch Datenbanken oder ähnliche Einrichtungen, vorbehalten.

1. Auflage 2010
Copyright © 2010 GRIN Verlag GmbH
http://www.grin.com
Druck und Bindung: Books on Demand GmbH, Norderstedt Germany
ISBN 978-3-656-32434-8

Research Article

Effect of insect attack on the production of Honey & Honey bees in Nowshera District.

Alam Zeb, Dr Mohammad Amin (Shaheed Benazir Bhutto University Sheringal Dir Upper kpk Pakistan) & Dr Rajpar Nawaz (PFI Peshawar).

Abstract

The study was carried out in District Nowshera at 10 different locations namely; Nizampur, Akora khattak, Pirsabaq, Khat kaly, Azaa kheel, Chirrat, Khweshgi, Behraam kaly, Rashakai, Pabbi, and Kaka sahib through questionnaire in order to gather the information from beekeepers to investigate the predators and parasites of honey bee that causes the losses of honey production and on honey bees. The results show that mites caused severe loss of honey production (35.21%) followed by bee-eaters (21.69%), wax moth (21.63%), black aunts (14.79%) and hornets (6.77%) respectively in the study area. This research was conducted to determine the bee predators and parasites that causes the loss of honey production and honey bees in Nowshera District, Khyber Pukhtunkhwa.

Corresponding Author: Zeb.Alam

INTRODUCTON

General Background

Beekeeping or apiculture is the science of rearing honey bees in hives under semi-natural conditions for the production of honey and other useful products. A location where honey bee colonies are kept is called Apiary. Honey bees are also used by farmers to pollinate their crops, particularly oil seed and fruit crops.

Apiculture, once considered a cottage occupation, has now emerged a profitable industry in the world. Europe, America, Australia, China, and Japan produce a large quantity of honey and other bee products such as pollen, royal jelly, wax, portfolios and bee venom that have been export worldwide and earn handsome foreign exchange. Many countries imports queens from Australia, because these queens produced bees that are gentle in nature and produced high yield (Ruben, 2006).

Pakistan has ideal climatic conditions for beekeeping in different zones indicating bright chances for the development of beekeeping. Khyber Pukhtonkhwa is the hub of beekeeping due to the presence of important bee flora yielding high quality honey of shain, ber and phullai. Shain honey is produced in Swat valley and has higher medicinal value. Ber honey is produced in Kohat, Karak and Bannu, and exported to Saudi Arabia and United Arab Emirates due to high economic price. Similarly, phullai honey is produced in Chirat, Khairabad and Nizampur hills and used locally (Steven, 2008).

Importance of Beekeeping

Beekeeping has many uses; it produces food in the form of honey. It assists in the production of food by increasing yield of fruit and seed crops. It produces basic raw materials, beeswax, for which there are over 200 different uses. It helps in the regeneration of the forest, the reclamation of eroded land and the improvement of pasture. It gives a profitable and healthy form of livelihood to large number of people, and it is of considerable importance in the economy of the country (Pourta, 2009).

Systematic position of Honey Bee

Honey bees belong to:

Phylum: Arthropoda

Class: Insecta

Order: Hymenoptera

Family: Aphidea

Genus: Apis

Species: sp.

Morphology of Honey Bee

The body of honey bee is divided in to head, thorax and abdomen. It has a pair of antennae, three pairs of legs and two pairs of wings (Figure 1).

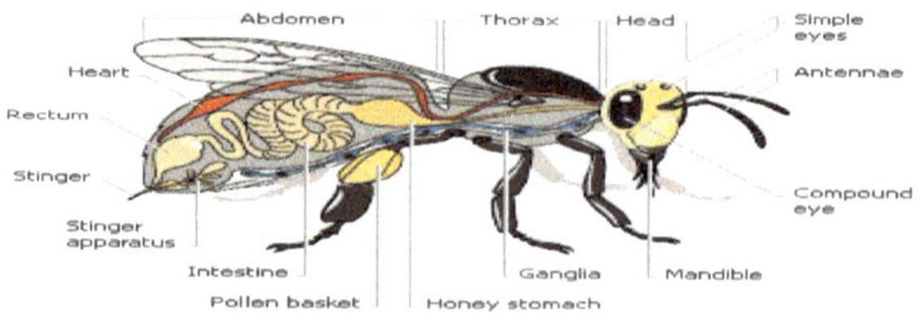

Source: http://www.insct-morphology.seql.eda.bee.html

Honey Bee Species Occurs in Pakistan

Four species of honey bee occurs in Pakistan, which as as under;

Apis dorsata F. (Rock bee)

This species is of the large size and produces plenty of honey. It constructs a single large combe, sometimes four feet long, on tall forest and shade trees, walls, parts of buildings and in rocks etc. it is unfit for domestication because of its peculiar habit of single combe building, migrations and bad temper.

Apis florea F. (Small bee)

This is the smallest species of all and is often called the little bee. It builds single combe in bushes and hidden places and cannot de domesticated. The combe is small is size and produces littgle honey.

Apis cerana indica F. (Indian bee)

It is small in size than *A. dorsata* and *A. mellifera* but larger than *A. florea*. Members of this speices produce combs in parallel rows inside natural hollows in the trees, walls, almirahs and other similar protected places. Honey yield from this species is higher and the bees are eomaratively miled in temper. It is found in the plains and mountains. This is the species that has successively been domesticated and is normally reared in most of the apiaries.

Apis mellifera F. (European bee)

It is successively reared in hives in Europe, America, Australia and Africa. A number of strains of this bee exist in various continents. The italian variety is quite common. It has been introduced in Pakistan, India, China, Japan, and many other countries. It is a bit larger in size than *A. cerana* and is high in honey production.

Casts of honey bees

Honey bees are social insects and there is division of labors among them. An average colony of *Apis mellifera* consists of 50,000-80,000 bees and there are three kinds of individuals such as workers, drones and the queen.

i.   Workers:

These constitute more than 90% of the members of a colony. All the workers are sterile females and do not lay eggs. They are comparatively smaller in size than drones and queen in size (Figure 2). The main duties of the workers are gathering the nectar and pollen from flowers, building comb, rearing brood and defense colony and attending the queen. A worker can live from 6 weeks to 6 months (depending upon the weather and extent of labors).

ii.   Drone:

They are male bees, slightly larger in size than workers and smaller than queen. The abdomen is black in color and without a sting. Each colony consists of 500-1000 drones and sometimes they are entirely absent. The main duty of drone is to fertilize the queen. The mouth parts are reduced and they feed on honey collected by workers (Figure 3). The life span of drone ranges from 2 to 4 months, depending upon the weather and extent of labors.

iii.   Queen:

Normally a colony has only one queen. The important characteristic of queen is large extendable abdomen beyond the wings. The queen is long in size as compare to other individual of the colony (Figure 4). She has a large conspicuous sting for killing other queens and keeps the colony strong. She leaves the hive only at swarming time or during mating. Mating takes place in the air. Queen laid 2000 eggs in 24 hours under favorable conditions, and can live for 3-5 years. A queen lays fertilized and unfertilized eggs. Fertilized eggs hatch into workers and queen and the unfertilized eggs into drones only.

LIFE STAGES OF HONEY BEES

Honey bee eggs hatched into young larvae that feed on pollen and honey collected by workers in the form of "honey combical wax". When larvae gain full size, they pupate. From pupae the adult emerge. The durations of the each stage in life cycle of honey bee may vary (Table 1).

Table 1: Different life stages of honey Bee

Caste	Egg Stage	Larval stage	Pupal stage	Complete cycle
Worker	3 days	6 days	12 days	21 days
Drone	3 days	6 ½ days	14 ½ days	24 days
Queen	3 days	5 ½ days	7 ½ days	16 days

PREDATORS OF HONEY BEES

Honey bees are social insects, however, these insects preyed by different insects such as wax moth, bee-eater, hornets, Mites and Black Aunts. The details of each predator and parasites are given below;

Wax-moth: The caterpillars of moth make silken tunnels in the comb and feed on bee-wax. When attack is severe, the comb is covered with silken tunnels with numerous black particles, the faucal matter of the caterpillars. In such cases, the bees abscond (Figure 5). The pest can be controlled by keeping the colonies strong and checking weekly to kill the larvae of Moth.

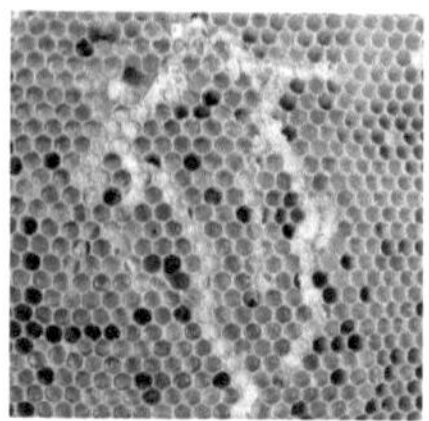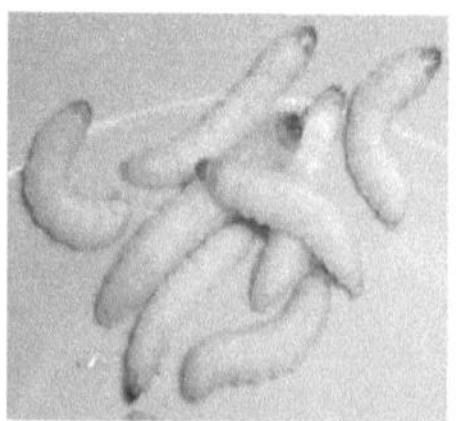

Hornets: Four different species of hornets destroy bees at the entrance of hive and in the field (Figure 6). Kill them with a fly flapper at the apiary. Locate their nest and destroy by insecticidal spray.

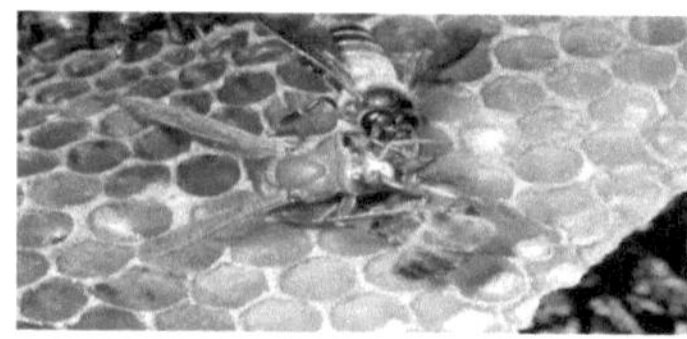

Black ants: They take away honey, brood and pollen. They fight with the bees, when infestation is high, the bees abscond (Figure 7). They can be controlled by placing the bee-hive on wooden stands with their legs in earthen cups containing water.

Bee-eater: These are the bird seen sitting in small parties on telegraph wires, poles and solitary trees. They caught bees through flying (Figure 8). They can be controlled by shooting.

Source: *http://en.wikipedia.org/wiki/File:European_bee_eater.jpg*

Mites: Mites attack the brood and adult bees usually in February and September. The infested brood produces bees with detective wings (Figure 9). In adults, the mites enter into the respiratory tracts and kill the bees. Mites can be controlled by employing sulphur dust or formic acid or folbex treatments.

Source: http://www.insect-beepredators.sqlm.fsl.html

MATERIAL & METHODS

Study area

The study has carried out in District Nowshera at 10 different sites / locations namely; Nizampur, Akora khattak, Pirsabaq, Khat kaly, Azaa kheel, Chirrat, Khweshgi, Behraam kaly, Rashakai, Pabbi, and Kaka sahib in order to gather the information through questionnaire from beekeepers to investigate the predators and parasites of honey bee that effect on the production of honey bees and their products. Nowshera is encompassing of forty seven union councils with an area 1743 km^2. This area falls under tropical thorn forest and vegetation comprises of wide array of tree species and the Bee flora in the whole District.

The below Map "Fig A" showing the study area of Nowshera district. (Fig. A)

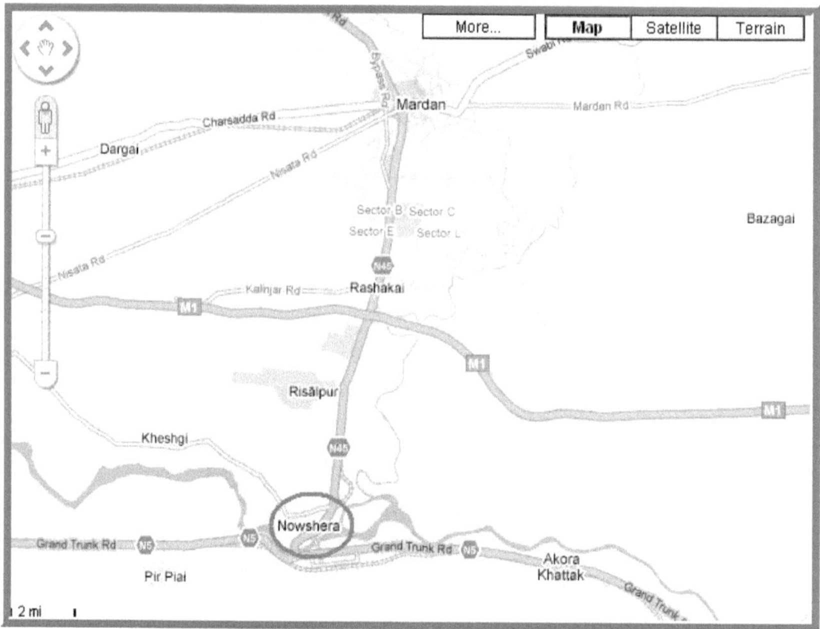

Fig A: Showing study area of Nowshera district

Study Survey

From the study, it was found that insect's attack's is causing issue in Nizampur. Insect's which are found in these form's are mite's, wax mouth and Bee eater , they cause the following losses shown in tabulated form respectively. In these forms it was studied and found that mite is present in large number and cause 30% loss in each form. In summer season, Bee eater was appear there and also cause 30% loss in production of Honey and Honey Bee. The insects found in Pirsabaq were Bee eater, Bee aunts and their losses were recorded as Mites30% to 80%, Wax mouth 70% to 90%, Bee eater 30% to 70%, and Bee aunts 60% to 90%. In Pirsabaq, mites and Bee eater are found in different ratios in different Bee forms. Mites was found in great number and caused effective loss in Honey and Honey Bee production in Khat kaly. Also Bee eater and Black Aunts cause effective loss in these forms . In Azaa Khel, Mites are found in great number and have caused 30% to 90% loss in Honey and Honey Bee Production. The most common insects in Chirat are Bee eater, Mites, wax moth but the exclusive loss was caused by Mites and Bee eater. The Mite caused 30%--to -90% losses to the production of Honey and Honey Bee. Mite cause 30% to 90% loss, Bee eater cause 20% to 50% and Bee aunts cause 10% to 60% in area of Khweshgi, However Mites are found in great number and cause exclusive loss. In Behram Kalu Rashakai, Bee eater and wax moth are found and cause 30% and 40% loss in the honey bee production and honey bee respectively. Like other areas Mites are present in this area, but in less number as compare to other areas, because the reason has told that when the number of honey bee is maximum / higher, there are no chances of Mites, but when bee are in less number there must be Mites present and vice versa. Main insects are Bee eater ,Wax moth and Mites but the great losses are caused by Bee eater in this area, 30%- to -90% loss are caused by Bee eater, and 10%-to-30% by other insects in the forms of Pabbi. In Kaka Sahib,, Mites are mostly found in great number in these forms because this insect is produced in winter and increasing in summer while summer season is also extendable in kaka sahib. Mites are effectively found in great number and cause 30 % to 90 % loss and other insects are less effective in Kaka sahib.

Data Analysis

A simple multiresidue method for the determination of insecticides in honeybees is described. The developed method is based on the matrix solid-phase dispersion technique(Emlen, 2011). The losses was calculated from well developed Questionnaires & Ex-Research study.

RESULTS AND DISCUSSIONS

In Nowshera, insect's effects were in different ratios i.e. Ranging from 10%-90% as a whole. Sometimes depend upon the seasonal changes because in winter the mites attacked and in summer this attack was increased. The mites were found in Kaka Sahib,

Khat kaly, Khweshgi, Chirrat, Azaa kheel, Akora Khattak, Pirsabaq and other areas of Nowshera except Behraam kaly (Rashakai Nowshera) and Ajab Baagh, Chuki Mumraiz (Pabbi), where bee-eater and black aunts were found in maximum number. The overall result shows that Mites causes honey loss of honey production 35.21%, wax moth 21.63%, Bee eater 21.69%, Black aunts 14.79% and the lowest loss was caused by Hornets as a whole 6.77 %. In addition it was founded that bee eater cause 32 % loss of honey bee that indirectly also cause on the production of honey

DISCUSSION

Mites are found in kaka sahib in large number because in kaka sahib the summer season is prolonged and the mites appear in winter; grow up in summer at high temperature. And lower in Rashakai Behraam kaly because the reason was told that when there is large number of bee in seasonal colony, so there must be least number of mites. (Alam, 2007). There are two varieties of moth which take delight in dining on wax the 'Greater' and also the 'Lesser' Wax Moth the greater wax moth is a mottled grey in colour approx 1 -1B= inches in length while the lesser is smaller and slimmer approx B= inch in length and white/silver. As all moths, they prefer night time to mate and lay eggs. (Photos are available in Introduction chapter).Most wax moths are seen in early summer in our area, and we see them under the overhang of hive roofs, out of the daylight, when the hive is disturbed they take off quickly and disappear into the trees.

Preferring to work in the dark the moths enter the hive through top entrances left unscreened and unguarded by the bees, perhaps a sudden cold snap making the bees cluster, and lay eggs in cracks unavailable to the bees. These hatch in due course and the grey larvae begin feeding on wax and hive debris, tunneling just under the cell caps and feeding on the discarded cocoons left by the bees, leaving behind an extremely sticky white web, similar to spiders web but almost impossible to pull apart. So perhaps they are misnamed and should be called Cocoon moths. They cause the lowest cause in the production of Honey & Honey Bee (NIFA ,2008).

Bee-eaters occur was found in large number in Pabbi and Choki Mumraiz as the environment of these areas is suitable for the production of Bee-eater. That's why the losses due to bee-eater was in large value in Pabbi and Chuki Mumraiz and lowest in other areas of Nowshera. Wax moths are found in moderate number in all forms due utilizing insecticides recently introduced by Tarnaab form Peshawar with proper training to control wax moth that causes loss on honey bee production. Less loss was recorded due to hornets because during winter they hibernate and less active. In addition, bee-keepers also employ insecticides to control them. Almost in all the study area, black aunts are also found in large number in summer and least in winter because all the insects are going for hibernation in winter, and their effects are sever in summer due to high temperature (Larry, 2010).

Bee-eaters predominantly eat flying insects, especially bees and wasps, which are caught in the air by sallies from an open perch. While they pursue any type of flying insect, honey bees predominate in their diet. Hymenoptera (ants, bees and wasps) comprise from 20% to 96% of all insects eaten, with honey bees comprising approximately one-third of the Hymenoptera. Before eating its meal, a bee-eater removes the sting by repeatedly hitting and rubbing the insect on a hard surface. During this process, pressure is applied to the insect thereby extracting most of the venom. Notably, the birds only catch prey that are on the wing and ignore flying insects once they land. Bee-eaters are gregarious. They form colonies by nesting in burrows tunneled into the side of sandy banks, such as those that have collapsed on the edges of rivers. Their eggs are white and they generally produce 2-9 eggs per clutch (depending on species). As they live in colonies, large numbers of these holes are often seen together, white streaks from their accumulated droppings accentuating the entrances to the nests. Most of the species in the family are monogamous, and both parents care for the young, sometimes with the assistance of other birds in the colony. They cause moderate loss in the production of honey & honey bee ranging from 30 to 60 % as a whole (Houston, 2011).

Why the mites are low in Pabbi & Behram kaly Rashakai? Because the reason was told as: A simple multiresidue method for the determination of insecticides in honeybees is described. The developed method is based on the matrix solid-phase dispersion technique. A total number of 12 insecticides (azinfos-methyl, buprofezin, chlorpyriphos, chlorpyriphos-methyl, diazinon, ethion, fenitrothion, fipronil, methidathion, phosalone, pirimicarb, propoxur) used on flowering fields are determined by this method. The method uses Florisil and silica as dispersing agents, alumina and silica as cleanup

adsorbents and a low polarity solvent system to elute pesticide residues from the honeybee samples. The insecticides were quantified using capillary gas chromatography with a nitrogen-phosphorus detector. The method has shown good recovery (70-110%) for various levels of spiked samples (0.01-1.0 mg/kg). The relative standard deviations were in the range of 2-8% for all pesticides studied. The limits of detection were in the range of 0.005-0.05 mg/kg. The procedure can be applied for the determination of residues of low-polarity and medium polarity pesticides in honeybee samples.
(Emlen, 2011).

Why after flood the mites have caused more loss in the production of honey in Nowshera & other areas? Because the report given by **PROF(DR) SURAYYA KHATOON (Chairperson, Department of Botany, University of Karachi), that** Millions of spiders have taken refuge in trees due to the floodwater, which is still standing in many parts of Nowshera. It is stated that the local people have never seen anything like this before, but they claim to notice fewer mosquitoes because spiders are supposedly capturing them. Apparently an expert has not examined these trees and the infesting organism. Reportedly thousands of trees have been affected. Anything with certainty can be said only after a close examination, but I highly doubt that these are spiders. The spiders using a tree just as an abode never smother it in their web in a manner like this. The typical look of this web on the trees in Nowshera is like a heavy infestation by spider mites. Although none of the pictures available on the Net is a close-up, all the affected trees (except one) are most probably acacia nilotica trees. The one which I could not identify looks like Salvadora or Zizyphus.Spider mites are not spiders. They are web-spinning mites parasitising on a number of plant species. They are serious pests which obtain their food by puncturing the plant cells; a heavy infestation ultimately kills the plant. There are more than 1,600 species of spider mites the world over, some of which may look like tiny spiders while some others may be hardly visible to the naked eye. Mostly they infest herbaceous plants, but several species parasitise on trees and shrubs. They spin the web not for capturing insects but for protecting their own colonies against any predator. The plants under stress are more susceptible to spider mites. The trees surrounded by floodwater are surely under stress due to over watering. It is not clear how many trees on how much area are affected, and how many species of trees and other plants are sufferers.

However, acacia nilotica appears to be the main victim. If this is, in fact, a hitherto unrecorded species of spider mite in our area, it is an alarming situation. The smothered trees would die in a matter of months if the infestation continues as such. Acacia nilotica is regarded as the 'golden tree of Sindh' which, being an indigenous tree in complete harmony with the local environment, has innumerable ecological and economic benefits. Large-scale death of these trees will be a great damage to the environment and livelihood of people. The Forest Department and Plant Protection Department should pay immediate attention to examine these trees and identify and control the infesting organism (Surayya, 2011).

CONCLUSSION AND RCOMMENDATIONS

Based on the results mostly the losses of honey are caused due to Mites, lowest by Black aunts & Hornets and moderate loss caused due to wax moth & Bee eater in Nowshera District.

The following steps should be taken to improve beekeeping industry in the country and Reduce/Avoid insect attack.

1) Bee hives, appliances, tools and bees should be made available to the interested people at reasonable prices.

2) A small industry should be setup for manufacturing bee hives and other bee keeping appliances and tools.

3) Facilities should be provided to the concerned agencies to produce mass nucleus colonies and queens to be readily available to the people.

4) Intensive studies on bee flora should be carried out throughout the country.

5) Major sources of bee flora particularly trees should be introduced to all over the country to fill the gaps. Eucalyptus species being major sources and having longer blooming period can be planted every where.

6) Regular training programs for traditional bee keepers, interested farmers and beginners should be introduced to popularize the industry on modern lines.

7) Elaborate arrangements for survey and study the diseases and pests be made to avoid large scale epidemics as it happened in 1982 and 1983.

8) The existing research facilities should be strengthened to take care of all national and regional problems faced by bee keepers.

9) Suitable extension service backed by research accomplishments is created for motivating farmers and villagers to take up bee keeping as profitable industry.

Acknowledgments

No Achievements is possible without the will & assent of God. His mercy & bounties are countless and whole life is too short for thanking Allah's limitless help and love. The words are too small to thank his numerous mercies.

I am very thankful to all the staff members of the Forestry Department & to Dr. Moh. Rajpar Nawaz Lecturer in Pakistan Forest Institute **(PFI)** Peshawar.

I have no words to thank my advisor Mohammad Amin Lecturer in Forestry Department, University Of Malakand, For his keen Interest & guidance in this Research study.Last but not the least no acknowledgement could ever express the intensity of my emotional belonging to my parents. They will always serve as a beacon to enlighten my intellect.

REFERENCES

Ahmad.R (1981). A guide to Honeybee Management in Pakistan NARC, Islamabad. Lahore Razi publishers. Pp 3- 5.

Alam, F. (2011): *The* Bee *Hive: the Fact of the Honeybee*, Kaka sahib Nowshera, 03469137914.

Dakpoora, M. (2009). The taxonomy of recent honey been. Journal of Hymenopter Research .Pp 165 – 196.

Emlen, S.T.; P.H. Wrege (1996). "Forced copulations and intra-specific parasitism: two costs of social living in the white-fronted bee-eater". *Ethology*, 71 (1): 2–29.

Jan. S (2010). *Some Honeybee Flora for Apis spp in Pakistan*. Pak. J. Zool. 3: (1) Pp 201-218.

Kenneth A.C (2006). The Principles of Beekeeping Backwards. www.zoologyjoneuro-honeybee/uk.xm.exe

Laghari. N (2010). *Study of bee insects*, http://www.stn.isbn-laghari/res.3421/vio.com

Layman. C (2010). *Plight of the Bee - The Ballad of Man and Bee*. Viovio. pp. 82. ISBN 978-0615251332. http://www.viovio.com/shop/26787.

Naz. M and Ahmad. R. (2004). Some Honeybee Flora for Apis spp in Pakistan. Pak. J. Zool. Pp 3 -23.

Rehman, W and M. I. Chaudhary (2010). Bee Foraging plants at Nowshera. 35 (2): 71-75.

Rehman. A. (2007). Insect Attack on bee flora. www.wikipedia/insect-florae,untl.com

Saeed.N (2005) Study of bee flora, M.Phil. Thesis. Pp 11-21.

Sheppard, A.M. and Walter, S. (2008). Phylogenetic relationships of honey bees (Hymenoptera: Apinae: Apini) http://dx.doi.org/10.1016/j.ympev.2006.02.002

SUPPLEMENTS

Supplement No 1: <u>Bee flora in Nowshera:</u>

Botanical Name	**Local Name**
Eucalyptus camaldulensis	Lachi
Acacia modesta	Palosa
Trifolium resupinatun	Shaftal
Brassica campestris	Sarsoon
Zea mays	Jawar
Citrus sinensis	Malta
Psidium guyava	Amrood
Robinia psedoacacia	Kekhar
Melia azedarachi	Bekayana
Pisum sativum	Matar
Lathyrus odoratus	Sweet Matar
Berbaries lyceum	Tor meva
Dodonaea viscose	Ghorasky
Medicago polymorpha	Peshtaray
Indegofera sperma	Aam jamo
Syzigium cuminii	Jaman
Brassica oleracea	white sarsoon

Supplement No 2:

ANOVA TABLE:

ANOVA

Source of Variation	SS	df	MS	F	P-value	F crit
Sample	2001.316	1	2001.316	16.98629	0.000197	4.098172
Columns	9068.425	18	503.8014	4.276045	8.1E-05	1.882603
Interaction	1588.604	18	88.25579	0.749076	0.740828	1.882603
Within	4477.14	38	117.8195			
Total	17135.49	75				